THE MATRIX OF LIFE

The Matrix of Life
Copyright © 2023 by J. C. Collins

Published in the United States of America
ISBN   Paperback: 979-8-89091-061-5
ISBN   Hardback: 979-8-89091-062-2
ISBN   eBook: 979-8-89091-063-9

All rights reserved. No part of this publication may be reproduced, stored in a retrieval system or transmitted in any way by any means, electronic, mechanical, photocopy, recording or otherwise without the prior permission of the author except as provided by USA copyright law.

The opinions expressed by the author are not necessarily those of ReadersMagnet, LLC.

ReadersMagnet, LLC
10620 Treena Street, Suite 230 | San Diego, California, 92131 USA
1.619. 354. 2643 | www.readersmagnet.com

Book design copyright © 2023 by ReadersMagnet, LLC. All rights reserved.

Cover design by Kent Gabutin
Interior design by Daniel Lopez

THE MATRIX OF LIFE

A View of Natural Molecules from the

Perspective of Environmental Water

Second Color Edition

J. C. Collins

MOLECULAR PRESENTATIONS INC.

VALATIE, NY USA

THE AUTHOR

Dr. Collins received his degrees in chemistry from Wayne University and the University of Wisconsin. After employment at General Motors Research, E I Dupont Research and Sterling Winthrop Research Institute, he accepted a position as Associate Professor and Department Chairman at Illinois Wesleyan University. In 1967, he returned to the Institute to direct Medicinal Chemistry and Developmental Research until 1987 when he retired to devote full time to his driving interest in the role of water in providing spatial order in living cells.

He has a number of publications and patents to his credit and has had an organic chemistry reagent "The Collins Reagent" named after him. However, the study of spatial properties and surface hydration of natural molecules have been his major interest for the past fifty years. This small book provides a pictorial view of how the dynamic interchange between liquid and cubic-ice states of water on surfaces of vital molecules, not only provides direction for assembly, but accompanied by charge-coupling, their interactions with each other.

The book is designed to give those with even a limited understanding of chemistry and physics, a glimpse of three-dimensional images of natural molecules and their spatial relationships with surrounding water. Hopefully, everyone who reads this book may gain an appreciation for the amazing way natural molecules formed and evolved to give us life.

Illustrations were developed on Apple Macintosh and Dell computers using Adobe Illustrator™. Data for structural analyses and the production of drawings were obtained from the published literature. Custom physical model-building was performed primarily with Framework Molecular Model parts (Prentise Hall, Englewood Cliffs, NJ 07632).
First Edition: "The Matrix of Life" published 1991.
ISBN: 0-9629719-0-1
Library of Congress Catalog Card Number: 91-90379
Second Edition published 2023.
Supplemental information is included in
www.linearhydration.com
www.cubichydration.com

DEDICATED TO

My Wife Betty and our family for tolerating molecular models all over the house and the many hours I spent constructing, computing and writing.

ACKNOWLEDGEMENTS

Thanks are due to coworkers at Sterling Winthrop Research Institute and Illinois Wesleyan University who encouraged me to continue my study of water. After presentations at Scientific meetings, a few would say that the ideas were interesting - others said they were absurd.

The most encouraging comment came from Professor Linus Pauling who was sent a copy of the first edition of the book in 1991. He sent a note saying: "You are on the right track but the ideas are too simple."

In 2013, a preprint copy of the book "Biomolecular Evolution from Water to the Molecules of Life" was sent to Dr. Michael New, a Scientist in the Planetary Science Division of NASA. His comment was: "Your concept of Transient Linear Hydration and Quantized Hydration Patterning are valuable contributions to understanding the unique role of water in origin of life research."

Since water has been discovered on some planets, there has been an intense search for evidence of the molecules of life. Hopefully, this little book will assist in understanding how water may have been involved in their formation and function. Thanks to all those who have been involved in the search.

Table of Contents

The Prologue xv

The Transient Linear Hydration Hypothesis

Water and Cubic Ice 1
The TLH Hypothesis and Insulin Assembly 3
Linear and cubic Hydration of Receptor Sites 4
Hydration Patterning and Transfer RNA 6
Hydration Patterning of Proteins - Carboxypeptidase A ... 7
Biomembranes – Form and Function 10
Acetylcholine Receptor-Site 13
Human Dopamine Receptor-Site 14
Morphine, THC and Strychnine Receptor-Sites 15
Transient Linear Hydration of t-RNA and DNA 17

Evolution of the Molecules of Life

The Sugars – Ribose and Glucose 18
Starch – Critical Role in Molecular Evolution 20
The Nucleotides and Adenosine Triphosphate 23
The Ribonucleic Acids, Transfer-RNA and Ribosomes .. 24
The Amino-Acids and Polypeptides 25
The Deoxy-Ribonucleic Acids (DNAs) 26
Phospholipids and Biomembranes 27

References 39-37

THE PROLOGUE

How was it possible for the molecular debris which landed on the early earth to assemble spontaneously and produced living breathing creatures like you and me? Certainly, there must be some sort of "Law of Nature" which moved the small random molecules that landed on the earth into the extremely complex forms that exist today. And how is it possible for them to function in such an orderly way that we can be aware of our own existence?

We all know that physical objects don't spontaneously assemble themselves; they require a plan and the skill to put them together. If we are to believe the Second Law of Thermodynamics: all systems in equilibrium must move spontaneously from order to disorder, not the other way around.

In 1944, Erwin Schrodinger, one of the Fathers of Quantum Mechanics, concluded in his little book, *What is Life*, that it is liquid water which reversed the direction of molecular evolution[1] – lipid surfaces in water move spontaneously from randomness toward order - not as they do in air. He proposed that there must be some sort of order in water to have moved small random molecules into complex coordinated ones and that there might be some sort of a Law, in addition to thermodynamics, regulating space within living cells.

In 1957, Albert Szent Gyorgy who received a Nobel Prize for his study of muscle, in his little book Bioenergetics, indicated that, in moving from the resting state to the contracted state, there is a complete change in the structuring properties of water in the cell. As a summary of his studies, he concluded that "Water is the Mother and Matrix of Life"- it gave birth to natural molecules and provides the order/disorder properties for life.[2]

In the 60s, molecular orbital calculations suggested that the trimer, with three water molecules bonded together at 2.76 Angstroms (the same as in ice)[3] might be the most stable structural unit in liquid water,[4] and later that linear elements of five and six water molecules might form on lipid surfaces that do not bond with water.[5] Ten years later, Lumry and Rajender in a 100-page detailed thermodynamic study, concluded that as small molecules approach a protein, there is a unit energy exchange between enthalpy and entropy in water[6] - between the liquid and ice-bonding states.

When it was reported in an NMR study in 1973 that the water surrounding proteins and nucleic acids exhibits the peaks of ice, it was suggested that structure in surface water night be providing order in living cells.[7] But when protein structures were examined and no ordered water could be found, it was concluded that it is internal thermodynamics which regulates protein assembly.[8] However, based on the above studies, as well as extensive molecular model building, it appeared that short linear elements of hydration must be forming kinetically on lipid surfaces and between ions and extending out hexagonal and cubic patterning of water, just as it does when it freezes.

The ordered units must last about a million millionth of a second, too short a time to be identified. However, they must be there and they must occupy space, even though they may be difficult to detect and cannot be isolated. Like electrons which provide the order and regulate space in molecules, hydrating water molecules may do the same in living cells.

The concept seemed so reasonable that I published it in in 1991 and now am providing supporting evidence and applications to provide reasonable answers for difficult questions.

THE MATRIX OF LIFE

The Transient Linear Hydration (TLH) Hypothesis

Before we consider interactions between water and molecules, it is important to understand that there are two distinctively different modes of attachment between water molecules. In the liquid state, water molecules are held together by opposite charges on their

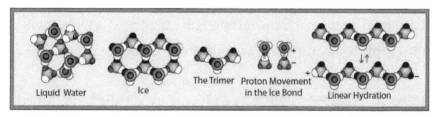
Liquid Water   Ice   The Trimer   Proton Movement in the Ice Bond   Linear Hydration

surfaces. The black electron orbitals on the molecules below have negative charges while the white orbitals have the proton of a hydrogen atom within them and a positive charge. Surface charges hold them together in the liquid state but they are in constant motion, jumping from one attachment to another; each lasting about a million millionth of a second.[9]

However, attachments between water molecules in the trimer and ice involve an overlap of electron orbitals in a covalent bond[10] with the proton nucleus of the hydrogen atom of one molecule next to the electron orbital of a neighboring molecule. The proton is so close to the neighboring water molecule that, by moving only a fraction of an Angstrom, it can positively-charge the adjacent water molecule.

When that happens in liquid water, hydroxide and hydronium ions are produced, but if it occurs in linear elements of hydration on the inner surfaces of nerve cells, positive charges can move through at almost super-conductive speeds.[11]

However, it is important to realize that water molecules in the liquid state are so dynamic and close together that they cannot spontaneously form ice-bonding at 0°C. Only if they are on a surface where the ends of lipid molecules or ions are in the same ordered positions as those in the surface of ice, can they freeze. If water is in a clean glass container, where silicon atoms are in random positions, cooling can be down to -30°C without freezing.[12]

At -40°C, freezing is spontaneous, but the ice produced is not the normal hexagonal form, it is the cubic isomer, which is produced most rapidly as orbitals overlap to form the more ridged linear covalent bond.[13]

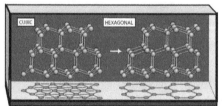

Cubic ice is called the "kinetic product" because it is the first molecular form produced as a new bond is formed. However, in water, the transition from the cubic form to the thermodynamically more stable hexagonal form occurs so rapidly that, even at 0°C, it cannot be detected.[15]

But, as Lumry and Rajender pointed out: covalent bonding between water molecules is unique.[6] As water molecules approach a molecular surface, either they form dynamic liquid-like attachments to polar oxygen or nitrogen atoms or they form short linear covalently-bonded linear elements on lipid surfaces. But, the elements are polar - they draw neighboring water molecules in line with them, withdraw a quantized unit of energy from the neighbor and extend linear elements out into an hexagonal pattern. Based on the conclusion of Lumry and Rajender that quantized units of energy are exchanged rapidly and repetitively, the process must generate transient ice-like hexagonal and cubic patterning on hydration-ordering surfaces.

Although the generation of this type of "kinetic order" is too rapid to detect at normal temperatures, Professor Zewail and his group at CIT, using ultra-high-speed crystallography at sub-zero temperatures, found that water on lipid and poly-ionic surfaces forms linear elements in cubic ice conformations.[16] Thus, the transient linear elements and cubic patterning generated on surfaces may, indeed, provide for cellular order. For example, when the insulin polypeptide is released from the ribosome,[17] nine peptides in the middle of the B-segment have side chains which shield the segment from binding water and force adjacent water to form unstable covalent linear elements on both sides.

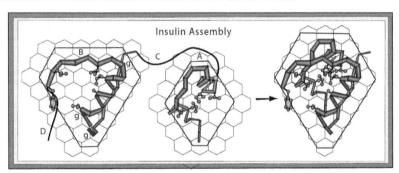

Insulin Assembly

The linear elements introduce so much instability into the B-segment that, as the water leaves, enough energy is withdrawn to convert it into the coil.[18] Two small glycine peptides (the gs at the lower end) bind directly with surface water, absorb energy and produce turns to direct the lipid front-side of the continuing segment up the back side of the coil. Glycine at the top of the coil directs the linear element, with ordering water on the back side, in a curve over to the left to position the ring of phenylalanine next to the valine methyls on the coil.

Segment C contains a number of small peptides which bind to water and remain mobile. C serves as a tether to direct the lipid front side of the A-segment into the back side of B. The A-segment forms two short coils with a short linear segment between them. By fitting into the lipid back side of B, unstable covalent water is lost from both surfaces. As C and D are removed enzymatically, the only region of the insulin protein that continues to be coated with unstable linear hydration is the lower right-hand side which binds into membrane receptors to activate processes in cells.[19] Notice, that its spatial structure mimics the cubic water which it displaces from binding sites in cell membrane.[20]

Just as the insulin molecule tends to mimic the spatial structure of cubic water and fit into receptor sites which control cellular functions, all hormone and neurotransmitter molecules examined to date appear to be spatial homologs of units in cubic ice.

As each of these vital molecules approach a receptor site, they must assume conformations which satisfy both their own internal thermodynamics, as well as the spatial requirements of water in low-energy linear an hexagonal forms in the sites. Although the

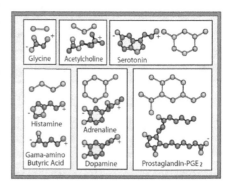

hydration forms produced in the site when empty last only about 10^{-10} second, they fill the space and hold it open for molecules that satisfy the binding requirements.

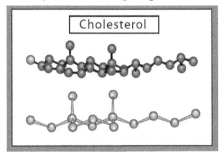

Cholesterol, which forms a complex with phospholipids to stabilize membranes, not only mimics the linearly ordered water but, by enzymatically losing its tail, is converted into a number of steroidal hormones which mimic six or seven linearly ordered water molecules.

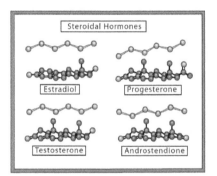
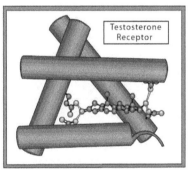

In 2006, Breton published the structure of testosterone in the binding site shown on the right above.[21] It is not hard to imagine how water, in the linear form shown on the left, might occupy the site, even for a short period of time as the testosterone molecule enters and leaves. In fact, when sites are not occupied by regulator molecules, they may open to admit water. The problem is that water does not occupy the site long enough to be isolated; it can be recorded spectroscopically, but not as a specific structure.[7]

Just as insulin and the small receptor molecules mimic cubically-ordered spatial units of water, small ribonucleic acids called transfer-

RNAs, which bind to multiple sites in large ribosomes to produce polypeptides, also mimic spatially-ordered units in cubic ice.[22] Details of the binding and polypeptide formation are included in www.linearhydration.com, but it is important to realize that there are at least twenty of these cubically-ordered t-RNA molecules in every living cell and the same number of enzymes with the structures shown below. The enzymes bind the specific three-letter nucleotide code on the left end of the t-RNA molecule to a sequence of peptides in the enzyme and attach a specific amino-acid to right-hand end.[23]

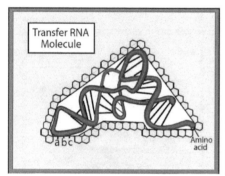
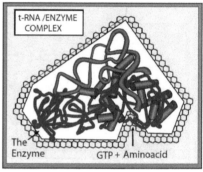

As you can see, the enzymes which bind these t-RNAs have complex structures, composed of both coils and linear segments - all fit together to form a structurally-stable anhydrous core. In this case, the t-RNA molecule not only binds to linear grooves in the enzyme, the enzyme itself reflects cubic hydration order with the end of the t-RNA molecule drawn down into the catalytic amino-acid binding site.[23]

However, small ions, like sodium and calcium, bind water molecules around them in spherical orientations.[24] Thus, their hydrated forms do not conform with linear elements of hydration which form on the lipid surfaces of proteins. Thus, some molecules mimic the spherical orientation of hydration around these ions. As nerve and muscle cells depolarize, internal water shifts from linear to spherical as sodium ions enter and calcium ions are released.[2]

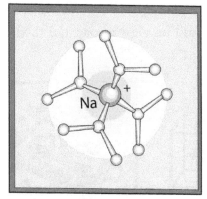

However, in spite of the spherical orientation of water molecules around these ions, when they pass through pores in membranes, they accommodate to the linearity of water molecules. For example, at the opening of a protein pore which passes through a membrane to admit sodium ions into nerve cells, there is an extremely toxic chemical which, during nerve-cell depolarization, binds within the pore and prevents sodium from entering. The chemical is tetrodotoxin. It is so toxic that, if consumed, causes almost immediate death.[25]

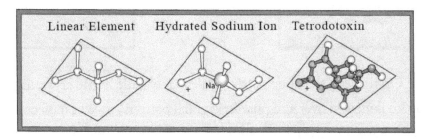

When the tetrodotoxin binding site is empty, water most likely fills the space. When a sodium ion enters, it jumps from one water-position to the next. However, when tetrodotoxin enters, it not only fills the site but binds so tightly that sodium ions cannot pass. Notice the interesting structure of the tetrodotoxin molecule: 7 oxygen atoms and 3 nitrogen's in an extremely small water-soluble molecule which mimics the structure of hydrated sodium.

When the transient linear hydration hypothesis is applied to proteins, the molecules must be oriented so they can be viewed parallel with major linear elements. For example, in the analysis of the crystallographic structure of the carboxypeptidase A enzyme,[26] the molecule was oriented so it could be viewed parallel to the terminal coil ending in peptide 307.

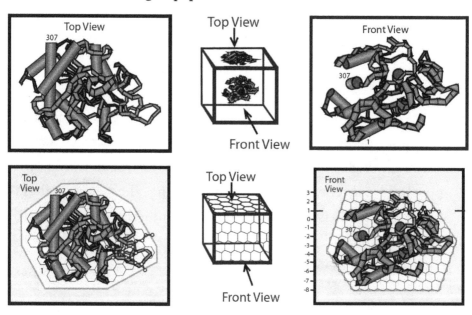

On the left, the upper surface of the molecule in the crystalline lattice is viewed over a regular hexagonal pattern of water molecules (2.25A apart in the plane). In the front view, with the cubic lattice shown behind, the coils and linear segments present an entirely different appearance - it is as if they are floating on planes of ice-like water with them numbered to define position. However, it must be remembered: even though the cubic lattice is illustrated as a complete unit, at any instant, only short linear elements are present and they last only about 10^{-11} second as the polypeptide folds and assembles to produce the anhydrous core. The matrix presented above is a cumulative view of the ordering units of surface hydration which propagate elements of order across lipid surfaces to guide assembly.

If we look at an enlarged view of the enzymatic binding site with the zinc ion at the base, we can see that the hydroxyl groups of the serine peptides at positions 157, 158, 159 and 162 are in spatial locations to direct transient linear elements of hydration out from the molecule to attract negative-ends of polypeptide chains in for cleavage.[27]

Of extreme importance, is the realization that, in the Transient Linear Hydration analyses of multiple water-soluble enzymes: *once cubic hydration patterning is established around the core of a protein, the patterning is maintained as a quantized unit, not only to direct assembly, but enzymatic function and lipid associations with other proteins.*[18]

Another important feature involved in the formation of the anhydrous cores of proteins is that lipid regions which fold together usually hold the same quantized lengths of linear elements of hydration so that, in forming the union, all water is lost between the segments to yield a stable anhydrous union.

For example, as the polypeptide of the carboxypeptidase enzyme is released from the ribosome, most of the 27 amino- acids on both sides of the chain at the 307 end have hydrocarbon side-chains which shield oxygen and nitrogen atoms in the chain from hydrogen-bonding with surface water. As ice-like bonding on those surfaces break and energy is transferred, the chain straightens and then rapidly forms the long coil.[28] A proline at position 288 produces a turn in the coil, and another 282, produces another turn. Oxygens at threonines 304 and 293 are in the proper positions to induce ice-like bonding of water next to the coil.

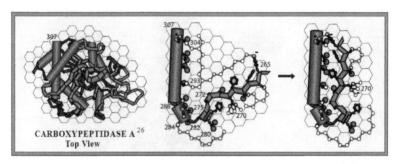

CARBOXYPEPTIDASE A [26]
Top View

From 280 to 275, the chain wraps back and forth in hydrated beta-turns so that the 272-265 segment can be at the level of the coil, rotate, bind and release equal lengths of unstable ordered-water from both surfaces.

Glutamic acid, at position 270, by clustering water around it to delocalize charge, breaks hydration order in the next hydration layer. At the same time, lipid surfaces above and below the acid continue to be covered by ordered water which, by being displaced by lipids, continues the formation of the anhydrous core. Polar and ionic side- chains of small peptides left on the surface, bond with surface water and other molecules at a variety of angles. Based on Transient Linear Hydration analyses, it appears that it is analogies in spatial-bonding properties between hydrated forms of natural molecules, and the order/disorder bonding properties of the aqueous environment, which permit the molecules to function in such a coordinated and spontaneous manner.

Now, if we look closely at biomembranes, we find that they are so loosely tied together that they cannot be isolated - they have to be constructed primarily based on spectroscopic data. In fact, it was in 1972 that Singer and Nicolson proposed the "Fluid Moseic Model" with the bilayer structure [29] and 1971 that Caspar and Kirschner published the electron-scattering (ES) and neutron-scattering (NS) curves for rabbit nerve cell lecithin/cholesterol membrane.[30]

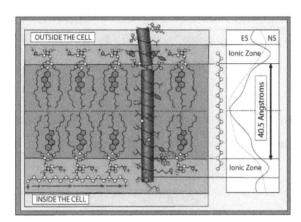 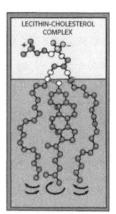

With the amino-phosphates of the phospholipids positioned at the peaks of the electron scattering curves on both sides of the membrane, (in the same locations as the ionic and polar groups on the protein coil which was isolated from a red-blood cell),[31] the lipid zone correlates to a distance of 40.5 Angstroms: 18 ice-like bonded water molecules (2.25Å per water molecule), lecithin/cholesterol molecules which meet in the middle and 27 peptides in the coil (1.5A between them). Once again, linear elements of hydration appear to have defined both the width of the membrane, as well as mean distances between phospholipids on the surface.

Furthermore, this idealized model provides an answer to a question that has plagued physiologists for years. In small nerves, when neurotransmitters like acetylcholine open pores in nerve endings to permit sodium ions into negatively-charged nerves, positively-charged potassium ions in the nerve carry the charge from end to end. However, in large axonal nerves, it is too far from end to end for potassium ions to carry the charge. In fact, the charge passes through axonal nerves at a much higher speed than small nerves.

The difference is that the inner walls of axons are composed, almost entirely, of lecithin/cholesterol complexes.[34] Positive charges generated in the nerve ending aline the polar heads of the lecithin molecules and, as illustrated in the figure above, positively-charged protons in adjacent ice-like-bonded linear elements of hydration carry the charge at almost super- conductive speeds from anionic phosphate to phosphate and amplifying node-to-node to the end.[33] For many years, molecular biologists have searched for a mechanism by which the positive pulse could be carried with very little loss of energy. Based on the Transient Linear Hydration hypothesis, the answer is through hydration elements which last about 10^{-11} seconds.

Most functional proteins in membranes, as illustrated schematically below, are composed of multiple coils which either rotate a single coil to signal a change in activity within the cell or rotate coils on both sides of a pore to admit and release ions and molecules into and out of a cell. In a third type, the central portion of a protein rotates to bring specific molecules in and out.[34]

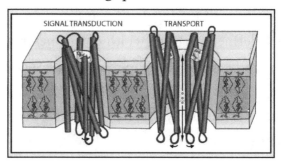

The Matrix of Life

When a neurotransmitter, like acetylcholine, binds to receptor sites on both sides of a transport pore, it opens to admit sodium ions to trigger the depolarization of the nerve cell.

Although a number of neurotransmitters open ion pores, acetylcholine is the most important in nerve and muscle cells. Notice that it is one of the smallest in mimicking the trimer of water.

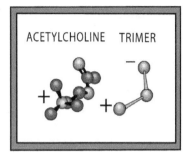

After an intensive study of the receptor protein in the electric eel, Dr. Nigel Unwin, in 1999, published the structure of the receptor protein and how it functions.[35]

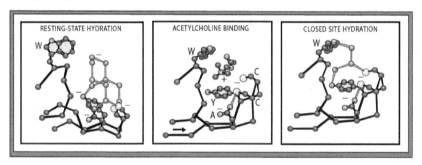

Based on the Transient Linear Hydration hypothesis, the receptor site most likely is highly hydrated in the resting state with linear elements of hydration periodically forming in preferred cubic locations. However, when the acetylcholine molecule enters, water is displaced and the two aromatic rings and two yellow sulfur atoms are drawn around it as shown above. When that happens, a polypeptide chain which is attached to a central coil, rotates the coil. As water enters to displace acetylcholine from the site, as shown on the right, it fills with a trimer of water and then opens to admit more water. Once again, it appears that water provides quantized spatial order for the spontaneous function of this receptor.

Since receptor sites in their active forms are extremely difficult to isolate, the TLH hypothesis is valuable in being able to view receptor molecules schematically in their activating forms. In fact, large "antagonist" molecules are often bound into sites in their resting states to provide stabilization and permit isolation. For example, in 2010, E. T, H. Chien and a group at the Scripps Institute, by changing several amino-acids in the coils and using a large antagonist molecule, were able to obtain a structure for the resting-state human receptor site for the central nervous-system "agonist" molecule dopamine.[36] Dopamine is a critical receptor molecule in the brain; if depleted, causes Parkinsons Disease.

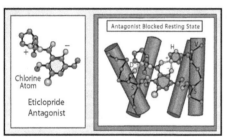

As illustrated, the eticlopride molecule essentially fills the binding site in its open form. Couplings with aspartate acid (D) behind the coil on the left and the histadine ring (H) on the right provide major binding. A serine (S) on the left bonds with an oxygen atom in the molecule while serine (S) on the V-coil is pushed out of position. Although dopamine activation is too labile to isolate, the TLH hypothesis permits the development of a schematic model.

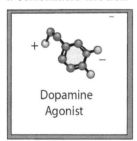

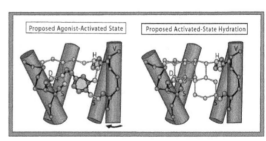

As mentioned in the introduction, cubic hydration patterning is generated kinetically by the transfer of a quantized unit of energy from one water molecule to the next.[6]

As might be expected, transduction receptor sites vary tremendously in structure based on binding peptides but most have similar coil structures and activating mechanisms. For example, the receptor sites derived for morphine[37] and tetrahydrocannabinol[38] are similar, but with substantially different binding peptides.

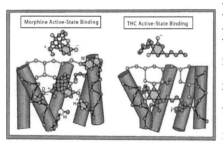

In the proposed figures on the left, the phenolic oxygen of morphine bonds to the cationic ring of histidine while the end of the tetrahydrocannabinol molecule rests on the aromatic ring of the phenylalanine. As the morphine and THC molecules leave the binding sites, water most likely assists in quantized steps.

Recently, a taste receptor site which binds strychnine was reported with the figure on the left below.[39] The molecule is flat and the view is directly down into the site from the surface. Once again, it is the cationic nitrogen in the molecule which binds to an anionic glutamate ion on the coil wall. Hexagonal TLH-patterning on the right fills the open space.

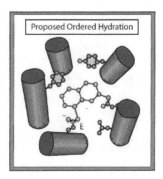

As more resting-state receptor sites become available, TLH Patterning will be applied to determine its value in illustrating how surface water most likely is involved in providing order, not only in receptor sites, but around all molecules in the living cell.

Last, but not least, we must consider DNA. Undoubtedly, it is the most well-known natural molecule but, at the same time, it is the most misunderstood because most people, including most in the scientific field, do not realize that DNA exhibits its uniform orderly structure only if surrounded by at least 13 ice-like covalently ordered water molecules.[40]

In 1953, when a number of research groups were searching for the structure of DNA, it was Rosalind Franklin, at Kings College in England, who sprayed a crystalline sample of DNA with water and obtained the pattern which Watson and Crick used to finish their model of DNA and publish the structure.[41] Unfortunately, the fact that Rosalind had obtained the pattern by spraying with water, was essentially lost in their publication.

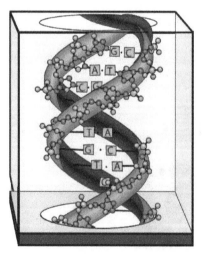
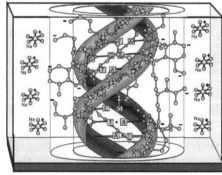

First: a linear element of hydration passes around the helix in the narrow minor groove.[42]

Second: short linear elements continually bridge between the strands,[43] and, last but not least: (based on model-building and the linear hydration hypothesis), water continually produces linear elements of hydration in cubic ice forms which span between phosphates in adjacent strands to delocalize negative charges and distribute them out to sodium and calcium ions around the helix." Although water is never shown around the helix[43] in the classical model of DNA, it is linear elements of hydration which fill the space and provide uniform spacing between the strands. However, (once again), the linear elements last for only about 10^{-10} second: they can be "seen" spectroscopically as an ice-like cage which surrounds the double helix but cannot be isolated or structurally identified.[7]

When the anionic phosphates of DNA bind to the cationic codes, or around spherical histone proteins, surface water is released and the strands separate, either to produce DNAs or RNAs or to provide for storage.

For many years, studies of living cells have focused on the molecular components, but, as Lumry and Rajender proposed in their 1970 thesis and as has been emphasized in this thesis, much more must be done to evaluate the thermodynamic and kinetic roles of water.[6]

In the first section of this little book, an attempt has been made to illustrate how the spatial structures of the major molecules in living cells tend to mimic units in cubic ice. The next section provides a view of how units in cubic ice may have been involved in the early formation and selection of the molecules of life.

EVOLUTION OF THE MOLECULES OF LIFE

Just as the molecules shown above bind to receptor and transport sites, there is a molecule that mimics hexagonal water which binds to multiple sites in every cell in the body.

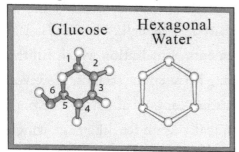

The molecule is glucose – the most abundant one on earth, with four oxygen's around it in the same spatial positions as water molecules in the hexagonal units in ice. The glucose molecule is produced by photosynthesis in plants and carries energy and carbon to every cell in the body to make almost every molecule in the body.

By mimicking the structure of hexagonal water, it either passes smoothly into cells or is transported there for energy and as a raw material. In fact, based on recent and old information, it may have been one of the first molecules to be made on earth.

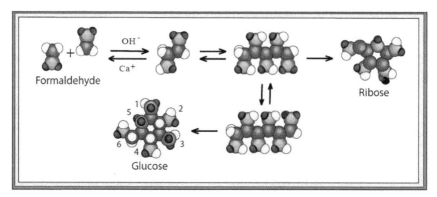

If formaldehyde molecules, which recently were found in the atmospheres of planets[44] (and are shown above with red oxygen's and white hydrogen's), are in weakly-basic aqueous solution, they bond together spontaneously to form five- and six-carbon units which circle around to form a variety of sugars, including glucose and ribose.[45]

Since glucose mimics the structure of hexagonal water, it bonds in numerous sites in cells. However, it differs from hexagonal water in one important respect: it binds water around it in a manner that differs from that of linear hydration bonding. It disrupts linear patterning of water around it by binding to water in liquid-like random dynamic fashion.[46]

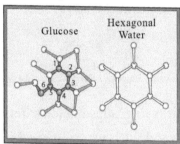
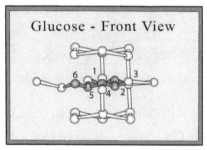

However, as illustrated on the right, alcoholic oxygens at positions 1 and 3 bond vertically with water molecules in linear and hexagonal patterning and those at 2 and 4, bond below.

Thus, glucose hydration, by being compatible with hexagonal order above and below but randomly distributed in the plane, gives the molecule the surfactant property of moving spontaneously into hydration-ordered surfaces, like membranes, to decrease order there, and then rapidly moving along the ordering surface in search of sites where it can move into cells.

As the most abundant natural molecule on earth, with the formula of $C_6H_{12}O_6$, glucose is both the spatial and carbon analogue of $H_{12}O_6$ water. As illustrated above, the arrangement of water molecules around it define how it moves and binds to other molecules. If it is accepted that glucose was one of the first molecules to form on the early earth, then it may have played a critical role in the formation of all other vital molecules. For example, if glucose is heated, its molecules combine chemically to form a complex mixture of polysaccharides, including the coiled strands of starch molecules.[47]

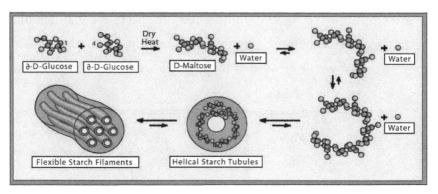

Although not shown, the OH (hydroxyl) groups, which extend out from the surface of the tubules, bond with water in random liquid-like fashion.[48] They solubilize and suspend the coils in water and permit them to bond with each other in filaments. However, the inner cores are hydrophobic, with the electron orbitals of oxygens directed toward the center.

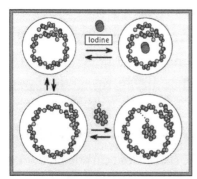

If we accept the thesis that glucose was produced in abundance from formaldehyde in the first phase of biomolecular evolution, then in the second phase, it was huge gelatinous masses of starch which most likely filled the oceans and tidal bays.[49]

Since starch coils have the property of binding small molecules, like iodine within their cores and unwinding to increase their diameter to bind larger molecules like hydroxynaphthalene,[50] it may have initiated the third phase of molecular evolution. Although the report which described that property did not include the nucleotide bases, they also may have bound within the cores.

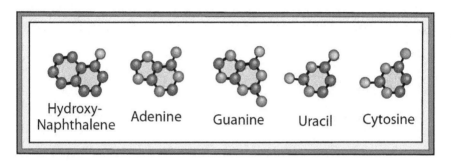

All four of the nucleoside-bases shown above have oxygen or nitrogen atoms in positions to bond with oxygen's in the walls to fill the cores. Of course, the ideas must be tested experimentally before acceptance. However, they may not only have bound

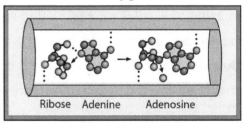

the bases, but a ribose molecule as well. On drying, nucleosides may have formed. Again, this is pure speculation and must be validated before acceptance.

Unfortunately, we have no way of knowing the conditions within the gelatinous masses which may have produced the nucleosides. The bases could have been produced in reactions between ammonia and formaldehyde, but nucleoside formation, if not as proposed, must have been performed by some other non-enzymatic process.

Once nucleosides began to be produced, the fourth phase must have involved phosphorylation. Although there were no enzymes to connect phosphate ions to the hydroxyl groups of glucose or the ribose rings of the nucleosides, polyphosphate ions, which can be produced simply by heating phosphate ions, form cyclic structures which have extremely high energy but are surprisingly stable in water. The polyphosphate ion might have bound to adenosine-sodium as shown below to form adenosine triphosphate (ATP).

Adenosine   Sodium-ATP

Again, we do not know, and never will know, exactly how these amazing molecules were formed in sufficient numbers to produce the complex systems which yielded life. However, ATP is one of most amazing vital molecules. With its triphosphate over the ribose ring, bound to the hydrated sodium ion, water molecules are held

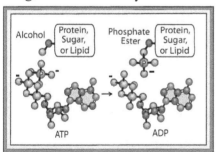

out away from the molecule, unable to bond with inner surface of the phosphorus atoms to break the bonds. The result is a sodium-ATP complex with sufficient hydration stability to move about in cells transferring its terminal phosphates to numerous other molecules. As sodium-ATP approaches a reaction site, a positive charge in the site displaces the sodium ion and binds the terminal phosphate in precisely the proper orientation, relative to the oxygen atom in another molecule, to transfer the phosphate to the other molecule. It gives the recipient molecule a negative charge and a leaving group.

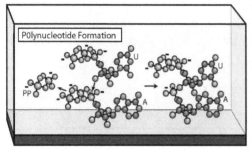

In addition to serving as sources of phosphate, nucleoside-triphosphates combined to form long strands of ribonucleic acids (RNAs) composed of adenosine, uridine, guanosine and cytosine, all tied together in a variety of sequences which would later provide the code for polypeptide synthesis.

Once tied together in RNA filaments, the nucleosides began forming specific attachments between adenosine/uridine (A/U) and guanosine/cytidine (G/C).

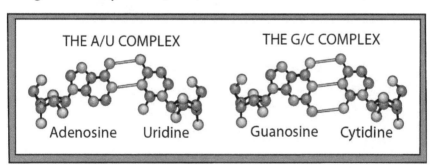

Regions of RNA which could couple A/U and G/C units together, circled each other to form linear double helices while other regions, as shown below, formed uncoupled turns and loops.

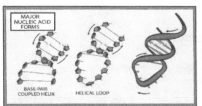
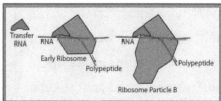

23

As noted on page 6, some of the first stable nucleic acids to form, most-likely, were the transfer RNAs. They had condensed anhydrous structures which corresponded to cubic patterning with ordered water and hydrated cautions around them. Although many of the RNA segments which formed at random were unstable and hydrolyzed back to nucleosides, some survived and produced an entire world of nucleic acids with the enzymatic properties of the proteins which were to follow. [55]

Based on correlations in structures, ribosomal particles may have grown from tRNAs with the capability of binding three t-RNAs in multiple orientations. At first, the ribosomes must have been small but, as they bound new RNAs and newly-formed proteins, they grew in size and functional specificity.[54]

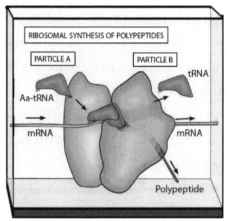

Obviously, we have no idea how those incredibly complex series of events occurred, but based on the structures of ribosomes and directions of channels through them, cubic hydration patterning most likely was involved.

As each of the amino-acids in the chart below became attached to the open end of a tRNA, (as shown on the right), with a specific sequence of three nucleotides on the loop, they were aligned next to each

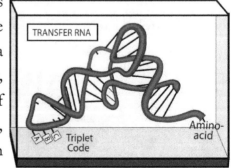

other on a messenger RNA and attached together to form specific sequences of the amino-acids, as shown above in polypeptides. Of extreme importance is the realization that studies have shown that water surrounds them as they are released from ribosomes and that some is lost as the polypeptides coil and wrap into the lower-energy protein structure. [18]

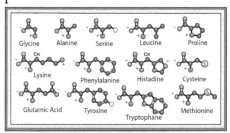

As we view the ribosome, tRNA the amino-acid molecules and the extremely complex processes evolved, we must wonder if there must have been a "Plan." Of course, we will never know the answer to that question and it will always be an article of faith.

When coded polypeptide synthesis first began, some of the earliest proteins to emerge most likely bound to open sites in the ribosomes to increase stability, productivity and specificity. Undoubtedly, many of them are still there. However, the production of enzymes, which could readily hydrolyze mRNAs back to nucleosides, threatened the entire coding process.

Only the formation of enzymes which could remove one of the hydroxyl groups from the ribose rings of the RNAs, was it possible for deoxy-ribonucleic acids to be produced which were stable enough in their double-helix forms, that enzymes were required to separate the strands.

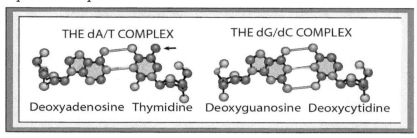

By replacing uridine with thymidine, (with a methyl group on the ring), a destabilizing water molecule in the RNA helix was displaced to increase structural stability. However, double-helix DNA was not only stabilized internally by water: a minimum of 13 water molecules were required, around and within it, to stabilize the molecule in its classically-viewed form.

However, once again, we must be reminded that most of the short linear elements which provide for the uniform repeating structure shown on page 16 last only about 10^{-10} second; they can be "seen" spectroscopically as an ice-like cage surrounding the double helix, but cannot be isolated.

Once genetically programmed enzymes began to be produced, simple molecules, like acetic acid, began to be assembled together to produce fatty acids and phospholipids of various lengths, with a variety of head-groups and a variety of charges.[56]

Like the hydrocarbon chains of the molecules in oil, when they contacted water, they align next to each other in layers based on lengths. The most abundant molecule that formed was a phospholipid named "lecithin" with fatty-acid chains of 18 carbons - particularly stearic acid, with a saturate chain and oleic acid with one double-bond.

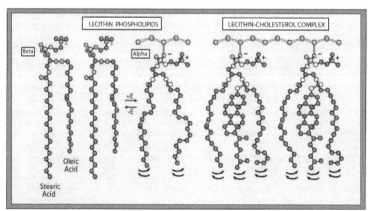

The Matrix of Life

At low temperatures, the chains lay side-by-side as shown above, but at normal temperatures, they have too much energy to remain straight, they twist and spin and occupy more space.[57] As shown above, the mean distance between phosphates in phospholipid/cholesterol complexes, which compose axonal myelin membranes of large nerve cells is about six linearly-bonded water molecules.

As shown on the right below, phospholipids form a double-layer with their dynamic tails in contact in the middle. Although they have a variety of head-groups, the major one in nerve and muscle cells has a trimethylamine group on the phosphate.[56] Proteins, with lipid surfaces of proper length on two or three sides, often assemble spontaneously in membranes to form pores.

Those with conical shapes and lipids on all sides, alter membrane to produce protrusions for binding with other membranes, while those with binding sites and pores provided for external and internal functions.

Once phospholipid membranes developed in evolution, functional units like ribosomes and mitochondria no longer had to be held as aggregates in gelatinous masses. Bacteria, molds, and viruses, which most likely were the earliest forms of life, combined in coordinated ways to bring forth living cells.

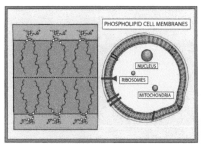

Although we will never know how living cells evolved, the laws of nature spontaneously brought it forth. It is incredibly complex, and yet, functions more smoothly and efficiently than anything man has made or may ever be capable of making. Let us hope that we have sufficient respect for what water and nature have done, that it will be here for many generations to come.

REFERENCES

1. E. Schrodinger, What is Life? with Mind and Matter. (*Cambridge University Press, 1944 and 1967*).

2. A. Szent-Gyorgi, Bioenergetics (*Academic Press, New York, 1957*).

3. S. N. Vinogradov and R. H. Linnell, Hydrogen Bonding (*Van Nostrand Reinhold, 1971*).

4. R. Hoyland and L. B. Kier, *Chim. Acta.* **15**: 1-11 (1969). Molecular orbital calculations for hydrogen-bonded forms of water. Also, J. Del Bene and J. A. Pople, *J. Chem. Phys.* **52**: 48-61 (1970). Theory of Molecular Interactions: Molecular Orbital Studies of Water.

5. C. Y. Lee, J. A. McCammon and P. J. Rossky, *J. Chem. Phys.* **80** (9): 4448 (1984). The structure of liquid water at extended hydrophobic surfaces. See also, L. F. Scatena, M. G. Brown and G. L. Richmond, *Science* **292**: 908-912 (2001). Water at hydrophobic surfaces: weak Hydrogen Bonding and Strong Orientational Effects.

6. R. Lumry and S. Rajender, *Biopolymers* **9**: 1125 -1227 (1970). Enthalpy entropy compensation phenomena in water solutions of proteins and small molecules: a ubiquitous property of water.

7. D. E, Woessner and B. S. Snowden, *Ann. NY Acad. Sci.* **204**: 113-124 (1973). A pulsed NMR study of the dynamics and ordering of water molecules at interfacial systems.

8. K. A. Dill and J. L. MacCallum, *Science* **338**: 1042-1045. The Protein - Folding Problem, 50 Years On.

9. A. Tokmakoff, *Science* **317**: 54 (2007). Shining light on the rapidly-evolving structure of water.

10. E. D. Isaacs, et al., *Physical Rev. Letters* **82**(3): 600 (1999). Covalency of the Hydrogen Bond in ice. A Direct X-Ray Measurement. See also, B. Dereka, Q. Yu, N.H.C. Lewis, W.G. Carpenter, J.M. Bowman and A. Tokmakoff, *Science* **371**(6525): 160-163 (2021). Crossover from hydrogen to chemical bonding.

11. O. F. Mohammed, D. Pines, J. Dreyer, E. Pines, and E. T. J. Nibbering, *Science* **310**: 83 (2005). Sequential Proton Transfer Through Bridges in Acid-Base Reactions. See also, A. J. Horsewell, *Science* **291**: 100 (2001). Evidence for coherent proton tunnelling in a hydrogen bond network.

12. H. S. Frank, *Science* **169**: 635 (1970). The Structure of Ordinary Water. See also, F. Franks (Ed.) Water - A Coomprehensive Treatise. (*Plenum, 1972*). Also, P. G. Kosolik and I. M. Svishchev, *Science* **265**: 1219 (1994). Spatial Structure in Liquid Water.

13. E. Mayer and A. Hallbrucker, *Nature* **317**: 601 (1987). Cubic ice form of liquid water. See also, A. K. Soper, *Science* **297**: 1288 (2002). Water and Ice.

14. P. Sykes, A Guidebook to Mechanism in Organic Chemistry, 6th ed. (*Pearson Prentice Hall, 1986*).

15. B. Kamb, Structural Chemistry and Molecular Biology pp. 507-542. (*Freeman, 1968*). Ice polymorphism and the structure of water.

16. D-S. Yang and A. H. Zewail, *Proc. Natl. Acad. Sci. USA* **106**(11): 4122-4126 (2009). Ordered water structure at hydrophobic graphite interfaces observed by 4D ultrafast electron crystallography. See also, C.-Y. Ruan, V.A. Lobastov, F. Vigliotti, S. Chen, A.H. Zewail, *Science* **304**: 80-84 (2004), Ultrafast Electron Chrystallography of Interfacial Water.

17. T. L. Blundell, J. F. Cutfield, S. M. Cutfield, E.K. Dodson, G.G. Dodson, G.G. Hodgkin, D.A. Mercola and M. Vijayan, *Nature* **231**: 506 (1971). Atomic positions in rhombahedral 2-zinc insulin crystals.

18. F. Mallamace, C. Cosaro, D. Mallamace, P. Baglion, H.E. Stanley and Sow-Hsin Chen, *J. Phys. Chem.* **B115**(48): 14280-14294 (2011). A Possible Role of Water in the Protein Folding Process. Also, R. L. Baldwin and G. D. Rose, *Proc. Natl. Acad. Sci. USA* **113**(44): 12462 (2016). How the hydrophobic factor drives protein folding.

19. C.W. Ward and M.C. Lawrance, *BioEssays* **31**(4): 422 (2009). Ligand-Induced activation of the insulin receptor.

20. For additional examples of the role of water in the assembly of water-soluble proteins see www.cubichydration.com.

21. R. Breton et al., *Protein Sci.* **15**(5): 987 (2006). Comparison of crystal structures of human androgen receptor ligand-bonding domain with various agonist-level molecular determinants responsible for binding affinity.

22. S.H. Kim, Advan. *Enzymol.* **246**: 279 (1978). Three-dimensional structure of transfer RNA and its functional implications.

23. M.A. Rould, J.J. Perona, D. Soll and T.A. Steitz, *Science* **246**: 1135-1142 (1989). Structure of E. coli glutamyl-tRNA synthtase complexed with tRNA and GTP.

24. F. Franks (Ed.) Water- A Comprehensive Treatise. (*Plenum, 1972*).

25. R. Chen and S.-H. Chung, *Biochemical and Physiological Research Communications* **146**: 370- 374 (2014). Mechanism of tetrodotoxin blocking and resistance in sodium channels.

26. F.A. Quiocho and W.N. Lipscomb, *Adv. in Protein Chem.* **25**: 1 (1971). Carboxypeptidase A.

27. W.N. Lipscomb, *Ann. Rev. Biochem.* **52**: 17 (1983). Structure and Catalysis of Carboxy- peptidase A.

28. K. Lindorff-Larsen, P. Stefano, R.O. Dror and D.E. Shaw, *Science* **334**: 517-520 (2011). How Fast-Folding Proteins Fold.

29. S.J. Singer and G.L. Nicolson, *Science* **175**: 720-731 (1972). The fluid mosaic model of cell membranes. See also, D. Chapman and D.F.H. Wallach (eds.) Biological Membranes. (*Academic Press, New York, 1973*). Also, D. E. Green (eds.) *Annual NY Acad. Sci.* **195** (1975). Biological Membrane Structure and Function.

30. D.L.D. Caspar and D. A. Kirschner. *Nature New Biology* **231**: 46 (1971). Myelin Membrane Structure at 10A Resolution. See also, D.E. Green (eds.) *Ann. N.Y. Acad. Sci.* **195** (1975) Biological Membrane Structure and Function. See also, D. Eisenberg, *Ann. Rev. Biochem.*

53: 595-623 (1984). Three-dimensional structure of membrane and surface proteins.

31. N. Mohandas and P.A. Gallagher, *Blood* **112**(10): 3939 (2008). Red Cell Membrane: Past, Present and Future.

32. F. de Meyer and B. Smit, *Proc. Natl. Acad. Sci. USA* **106**(10): 3654-3658 (2009). Effect of cholesterol on the structure of phospholipid bilayer.

33. A. Johnson and W. Winlow, *Physiology News* **38**(111): Summer 2018. Mysteries of the action potential - from 1952 to infinity and beyond.

34. M.S. Bretscher, *Sci. Amer.* **253**(4): 100-108 (1985). The Molecules of the Cell Membrane.

35. N. Unwin, *J. Mol. Biol.* **346**: 967 (2005). Refined structure of nicotinic Acetylcholine Receptor at 4A resolution.

36. E.Y.T. Chien, et al., *Science* **330**: 1091 (2010). Structure of the human dopamine D3 receptor in complex with a D2/D3 selective antagonist,

37. P. Nicolas, R. G. Hammonds and C.H. Li, *Proc. Natl. Acad. Sci. USA* **79**(7): 2191 (1982). Beta-endorphin opiate receptor binding activities of six naturally-occurring beta endorphin homologs studied using tritiated human hormone and naloxone as primary ligands.

38. W. DeVane, et al. *Science* **258**: 1946-1949 (1992). Isolation and structure of brain constituent that binds to the cannabinol receptor.

39. X. Weixiu et al. *Science* **377**(2022): 1298-1304 (2022). Structural basis for strychnine activation of human bitter taste receptor TAS2R46

40. P. Auffinger and E. Westhof, *J. Mol. Biol.* **268**: 118-136 (1997). Water and Ion-Binding around RNA and DNA.

41. J.D. Watson and F.H.C. Crick, *Nature* **171**: 737 (1953). Molecular structure of nucleic acid A structure of deoxyribonucleic acid. Also, J.D. Watson, The Double Helix (*The New American Library, 1968*).

42. M.L. McDermott, H. Verselous, S.A. Corcelli and P.B. Peterson, *ACS Cent. Sci.* **3**(7): 708-714 (2017). DNA's Chiral Spine of Hydration.

43. S.K. Pal, L. Zhao, T. Xia and A.H. Zewail, Proc. Natl. Acad. Sci. USA 140(24): 13746 (2003). Ultrafast Hydration of DNA. See also, S. Pal, P.K. Maiti and B. Bagchi, J. Phy.: Condens. Matter **17:S4317-S4331** (2005). Anisotropic and sub- diffusive water motion at the surface of DNA.

44. B. Zuckerman, et al., *Astrophys J.* **160**: 485 (1970). Observation of Interstellar Form- aldehyde.

45. A. Butlerow, *Justus Liebigs Annalen der Chemie*, **120**: 295 (1861). Formation of a sugary substance by synthesis. See also, C.R. Nollar, Chemistry of Organic Compounds (*W. B. Saunders Company, 1951*). p. 218. Note: The formation of sugars from formaldehyd has been questioned by L.E. Orgel, *Proc. Natl. Acad. Sci. USA* **97**(23): 12503-12507 (2000). However, ribose and other sugars have been isolated from primative meteorites, Y. Furukawa, *Proc. Natl. Acad. Sci. USA* **116**(49): 24440 (2019).

46. T. Suzuki, *Chem. Phys.* **10**(1): 96 (2008). The hydration of glucose: the local configurations in sugars hydrogen bonds. See also, J.C. Hower. Y. He and S. Jiang, *J. Chem. Phys.* **129**(21): 215101 (2008). A molecular simulation study of methylated and hydroxy sugar-based self-assembly into monolayers.

47. M. Vollameil, et al., 4 Browning Reactions (*Wiley-Blackwell, pp.83-85, 2006*).

48. N. Sharon, *Sci. Amer.* **245**(5): 90-93 (1980). Carbohydrates.

49. D.A. Rees, Polysaccharide Shapes (*Wiley, 1977*).

50. R.H. Marchessault and P.R. Sundararajan, *Adv. in Carb. Chem. and Biochem.* **33**: 387 (1976). Molecular Complexes in Starch.

51. G.E. Joyce, *Sci. Amer.* **260**: 90 (1990) Directed Molecular Evolution. See also, A. C. Wilson, *Sci. Amer.* **253**:164 (1985). The molecular basis of evolution.

52. O. Kennard et al., *Proc. Royal Soc. London* **325**: 401 (1971). The Crystal and Molecular Structure of Adenosine Triphosphate.

53. R.J.P. Williams, *Eur. J. Biochem.* **57**:135 (1975). Quantitative determination of the conformation of ATP in aqueous solution.

54. R. Benne and P. Sloof, *Biosystems* **21**(1):51- 68 (1987). Evolution of mitochondrial protein synthetic machinery.

55. T.R. Cech, *Sci. Amer.* **255**(5): 64-65 (1986). RNA as an enzyme.

56. G.L. Jandrasiak and J.C. Mendible, *Biochimica et Biophysica Acta.* **424**: 149 (1976). The Phospholipid Head Group Orientation: Effect on Hydration and Electrical Conductance.

57. J.L. Ranck, et al., *J. Mol. Biol.* **85**: 249 (1974). Order-disorder conformational transitions of hydrocarbon chains of lipids. Also, Kim et al., *J. Chem. Phys. Chem. B.* **110**(43): 21994 (2006). Ultrafast Hydration Dynamics in the Lipidic Cubic Phase. See also, G. Langley, *Angew. Chem. Int. Ed. Engl.* **15**: 575- 580 (1976). Kink-block and gauche-block structures of biomolecular films.